Cambridge Elements ≡

Elements in Geochemical Tracers in Earth System Science
edited by
Timothy Lyons
University of California
Alexandra Turchyn
University of Cambridge
Chris Reinhard
Georgia Institute of Technology

RECONSTRUCTING PRECAMBRIAN *PCO₂* AND *PO₂* USING PALEOSOLS

Nathan D. Sheldon
University of Michigan
Ria L. Mitchell
Swansea University and University of Sheffield
Rebecca M. Dzombak
University of Michigan

CAMBRIDGE
UNIVERSITY PRESS

CAMBRIDGE
UNIVERSITY PRESS

University Printing House, Cambridge CB2 8BS, United Kingdom

One Liberty Plaza, 20th Floor, New York, NY 10006, USA

477 Williamstown Road, Port Melbourne, VIC 3207, Australia

314–321, 3rd Floor, Plot 3, Splendor Forum, Jasola District Centre,
New Delhi – 110025, India

79 Anson Road, #06–04/06, Singapore 079906

Cambridge University Press is part of the University of Cambridge.

It furthers the University's mission by disseminating knowledge in the pursuit of
education, learning, and research at the highest international levels of excellence.

www.cambridge.org
Information on this title: www.cambridge.org/9781108819008
DOI: 10.1017/9781108870962

© Nathan D. Sheldon, Ria L. Mitchell and Rebecca M. Dzombak 2021

First published 2021

A catalogue record for this publication is available from the British Library.

ISBN 978-1-108-81900-8 Paperback
ISSN 2515-7027 (online)
ISSN 2515-6454 (print)

Additional resources for this publication at www.cambridge.org/sheldon

Reconstructing Precambrian pCO_2 and pO_2 Using Paleosols

Elements in Geochemical Tracers in Earth System Science

DOI: 10.1017/9781108870962
First published online: February 2021

Nathan D. Sheldon
University of Michigan

Ria L. Mitchell
Swansea University and University of Sheffield

Rebecca M. Dzombak
University of Michigan

Author for correspondence: Nathan D. Sheldon, nsheldon@umich.edu

Abstract: Paleosols formed in direct contact with the Earth's atmosphere, so they can record the composition of the atmosphere through weathering processes and products. Herein we critically review a variety of different approaches for reconstructing atmospheric O_2 and CO_2 over the past three billion years. Paleosols indicate relatively low CO_2 over that time, requiring additional greenhouse forcing to overcome the 'faint young Sun' paradox in the Archean and Mesoproterozoic, as well as low O_2 levels until the Neoproterozoic. Emerging techniques will revise the history of Earth's atmosphere further and may provide a window into atmospheric evolution on other planets.

Keywords: paleosols, oxygen, carbon dioxide, atmosphere, geochemistry

ISBNs: 9781108819008 (PB), 9781108870962 (OC)
ISSNs: 2515-7027 (online), 2515-6454 (print)

Contents

1 Introduction

The study of fossil soils (paleosols) began in 1726 with the first description of a buried Quaternary soil along the Danube River in Hungary, and many of the early foundational figures in geology such as James Hutton and Charles Lyell recognized that major unconformities represented erosional events (Retallack, 2013). However, paleosols were largely considered a curiosity within the realm of soil science or Earth science for almost 240 years after the initial description, before systematic efforts to recognize and document paleosols took off in the 1970s and 1980s. Paleosol-specific classification schemes and advances in stable isotope (e.g. Cerling, 1984) and whole-rock geochemistry (e.g. reviewed in Retallack, 1991) opened up new quantitative paleoclimatic reconstruction approaches, which in turn led to the compilation of extensive records of Phanerozoic paleosols (Sheldon and Tabor, 2009). Identifying Precambrian paleosols is more challenging (Retallack, 1992) because common Phanerozoic paleosol features such as root traces could not exist prior to the evolution of land plants in the Silurian (e.g. Kenrick and Crane, 1997), and many Precambrian weathering surfaces or paleosols had been deeply buried and subjected to metasomatism and/or metamorphism (Retallack, 1991).

Interest in paleosols as archives of the Earth's atmospheric composition in the Precambrian began in the 1980s. Because soils form at the Earth's surface, in direct contact with its atmosphere, they are potentially a simpler archive to interpret than marine deposits or chemical sediments such as banded iron formations (BIFs). Although some earlier workers drew qualitative conclusions about atmospheric oxygen from paleosols (Dimroth and Kennedy, 1976; Gay and Grandstaff, 1980), Heinrich (Dick) Holland was the first to propose explicitly that the chemical composition of paleosols could offer quantitative insights into the composition of the atmosphere under which they formed (Holland, 1984). The broader context for Holland's pioneering work was a community-wide debate on the amount of oxygen in the Precambrian atmosphere and the question of whether there had been a state-change (typically referred to as the Great Oxidation Event; GOE) between a nearly anoxic world and weakly oxygenated one. The Dimroth–Kimberley–Ohmoto model (DKO; Ohmoto, 1997) postulated that the Earth was oxygenated to modern levels early on its history, possibly in the Hadean, and the Cloud–Walker–Kasting–Holland model (CWKH; Holland, 1999) proposed a multistep oxygenation increase, with modern levels reached only in the Phanerozoic. At the centre of both sides of the argument was the oxidation state of Fe in Precambrian paleosols (Figure 1). Today, the broad debate has essentially been decided in favour of the CWKH model, although evidence for episodic 'whiffs' of oxygen prior to the GOE

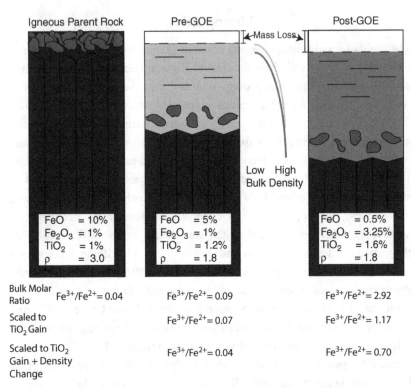

Figure 1 Schematic cartoons of pre- and post-GOE weathering. For both scenarios, exemplar data are given and in the table, changes in the Fe^{3+}/Fe^{2+} ratio are given for the bulk measured value, the ratio scaled to account for the immobile element Ti, and the ratio scaled for both Ti and that the bulk density of soil is less than that of bedrock. For the pre-GOE case, the Fe^{3+}/Fe^{2+} ratio appears to double relative to the parent material, but when mass balance and density change are accounted for, the value matches the parent rock. While the post-GOE Fe^{3+}/Fe^{2+} ratio has increased relative to the parent material, Fe^{2+} has not all been quantitatively retained in the soil given. If all of the parental Fe^{2+} had been oxidized to Fe^{3+}, then the 'fully oxidized' soil should have 6% Fe^{3+}. Thus, even though the Fe^{3+}/Fe^{2+} ratio indicates oxidation relative to the parent material, it fails to represent that Fe was still lost from the soil.

(Anbar et al., 2007) suggest that some of the arguments from the DKO model for evidence of occasional surface oxidation in the Archean were correct, even if their proposed levels of oxygen were never reached until the Phanerozoic.

While Holland was primarily interested in reconstructing pO_2, his interpretive framework for paleosols has led to a variety of approaches to reconstructing pCO_2 (e.g. Sheldon, 2006, 2013; Driese et al., 2011; Alfimova et al., 2014;

Kanzaki and Murakami, 2015) as well as more complex approaches to understanding pO_2 (Murakami et al., 2011; Yokota et al., 2013). Holland (1984; Holland and Zbinden, 1988; Pinto and Holland, 1988) proposed the following steady-state model for using paleosols to understand the balance between oxidation and carbon consumption (R) at the Earth's surface:

$$R = \frac{D_{O_2}}{D_{CO_2}} \sim \frac{0.25 m_{FeO} + 0.5 m_{MnO}}{2[m_{CaO} + m_{MgO} + m_{Na_2O} + m_{K_2O} + m_{CaMg(CO_3)_2}]} \tag{1}$$

where m_i values represent molar concentrations. D_{O_2} represents the O_2 required to oxidize all of the Fe^{2+} and Mn^{2+}, based upon the idea that they are the two most abundant reduced metals in most rocks. D_{CO_2} represents the carbonic acid required to remove all Ca, Mg, Na, and K, based upon the idea that they are the four most common rock-forming cations that are mobile during weathering (Sheldon and Tabor, 2009). Other rock-forming elements such as Al and Ti are essentially immobile during weathering, except under somewhat exceptional pH conditions (i.e. pH < 4 or pH > 9). Other elements like Si have complex weathering behaviours where they are immobile in some minerals (e.g. quartz) and highly mobile in others (e.g. feldspars). There are two key caveats to this approach: 1. post-burial alteration and diagenesis need to be considered, particularly with K, which is subject to metasomatic redistribution (Maynard, 1992; Driese et al., 2007); and 2. that Eq. (1) represents a case of 'one equation–two unknowns', meaning that either D_{O2} or D_{CO2} must be known or estimated to obtain the other value.

An example of this type of calculation, and of how it may be refined using more recent pCO_2 models can be found in Figure 2. Holland et al. (1989) estimated R for the ~1.9 Ga Flin Flon paleosol and, based upon various sensitivity tests, argued for a pO_2 ranging between 1.2×10^{-3} to 3×10^{-2} PAL (pre-Industrial atmospheric levels), with a best-guess value of 2×10^{-3} PAL, but noted that the pCO_2 could have varied over three orders of magnitude. Sheldon (2006) used a mass-balance model to estimate pCO_2 using the same dataset (Table 1), so by using that value and the calculated R value from Holland, it is possible to refine pO_2 estimates further (Figure 2). The revised estimates are both more precise relative to the earlier work that assumed pCO_2 without any quantitative constraints, and affirm results based upon marine proxies that suggest low pO_2 throughout the Proterozoic (Lyons et al., 2014; Planavsky et al., 2018).

These types of refinements to both estimates of pCO_2 and pO_2 based upon paleosols have implications for constraining temperature history and habitability of the early Earth, as well as providing guidance for what we might expect to find on other Solar System bodies or on exoplanets as we refine our ability to analyze

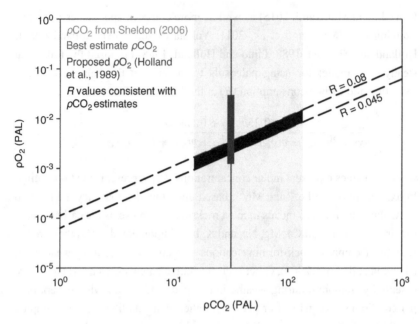

Figure 2 pCO_2 versus pO_2 for the 1.9 Ga Flin Flon paleosols. Holland et al. (1989) calculated R values (Eq. 1) for the Flin Flon paleosol or 0.045 (basalt parent) and 0.08 (greenstone parent) and used those values to calculate pO_2 ranging from 1.2×10^{-3} to 3×10^{-2}. Using pCO_2 estimates from Sheldon (2006) based upon the same dataset, we can refine estimated pO_2 further to 1.7×10^{-3} to 1.5×10^{-2}, which narrows the range and which affirms low Proterozoic pO_2 as proposed by various authors.

them remotely. Furthermore, quantitative estimates of pCO_2 and pO_2 have relevance for the geologic history of other greenhouse gases such as CH_4 or N_2O (e.g. Fiorella and Sheldon, 2017; Zhou et al., 2017). Changes in pO_2 are also linked to the history of seawater sulfate and to limits to biogeochemical cycling (e.g. Olson et al., 2016) in marine and terrestrial aquatic environments. For the rest of this Element, we will describe both previously proposed and currently used methods for reconstructing pCO_2 and pO_2 from paleosols, placing results from those methods into the broader Precambrian geologic context, examining some critical outstanding problems, and proposing several directions for future work.

2 Materials/Methods

Following the advent of land plants in the Silurian (Kenrick and Crane, 1997), one of the primary means for identifying paleosols is to look for root traces, which may range from kerogenized remains of the original roots to rhizoliths

Table 1 Strengths and weaknesses of different proxy approaches for pCO_2 and pO_2

Approach	Strengths	Weaknesses	References
pCO_2			
Steady state[1,4]	Relatively few free parameters; should work for incomplete profiles	Diagenesis; requires assuming pO_2 to calculate pCO_2	Holland (1984); Holland and Zbinden (1988); Pinto and Holland (1988); Zbinden et al. (1988); Rye and Holland (1998)
Mineral equilibrium[1,5]	Equilibrium K values well constrained for most minerals	Diagenetic alteration of mineral assemblage; assumes stoichiometric minerals; reliability of thermodynamic data; oversimplifies weathering	Rye et al. (1995); Hessler et al. (2004)
Mass balance[3,4]	Firmly grounded in soil science	Diagenesis; requires assumptions for some parameters (e.g. MAP, soil formation time); non-quantitative	Sheldon (2006; 2013); Mitchell and Sheldon (2010); Driese et al. (2011); Alfimova et al. (2014)[2]; Rybacki et al. (2016)
Dissolution kinetics[3,5]	Grounded in process-based understanding of weathering and mineral dissolution	Diagenesis; complex to apply; different kinetic assumptions give very different results; final answer is still basically a mineral equilibrium calculation and is subject to all of the issues associated with that method	Alfimova et al. (2014)[2]; Kanzaki and Murakami (2015; 2018a; 2018b)[3]

Table 1 (cont.)

Approach	Strengths	Weaknesses	References
pO2			
Steady state[1,4]	Relatively few free parameters; should work for incomplete profiles	Diagenesis; requires assuming $p\text{CO}_2$ to calculate $p\text{O}_2$	Holland (1984); Holland and Zbinden (1988); Pinto and Holland (1988); Zbinden et al. (1988); Rye and Holland (1998)
Fe(III)/Fe(II)[1]	*Fe minimally impacted by diagenesis*	*Non-unique interpretations; ignores basic soil science*	Ohmoto et al. (1996); Holland (1999); Utsunomiya et al. (2003)
Mass balance[1,4]	Firmly grounded in soil science	Diagenesis; requires assumptions for some parameters (e.g. MAP, soil formation time)	Driese (2004); Driese et al. (2007); Murakami et al. (2016)
Oxidation kinetics[3,5]	*Grounded in process-based understanding of weathering and mineral dissolution*	*Complex to apply; simplistic set of minerals considered*	Murakami et al. (2011); Yokota et al. (2013)
Cr isotopes[1,4]	Potential to measure both source and sink; clearly reflects surface conditions; cycle is linked other redox-sensitive metal Mn	Cr reactions with ligands; relies on understanding of Mn oxidation; potential for mobilization during high-temperature alteration	Crowe et al. (2013); Babechuk et al. (2017, 2019); Colwyn et al. (2019)

[1] Denotes model that can be considered qualitative or semi-quantitative
[2] Denotes forward modelling approach
[3] Denotes inverse modelling approach
[4] Denotes a model that relies on assuming a key unknown variable
[5] Denotes a model that relies on stoichiometric minerals and equilibrium

that represent mineral replacement of roots to drab-haloed root traces that reflect redoximorphic conditions during the oxidation of the original organic matter following burial. Other common features include colour changes, horizonation, burrows, pedogenic carbonates, and diagnostic micro-structures (e.g. gilgai micro-relief in Vertisols) that can be used to reconstruct pedogenic processes (e.g. Retallack, 1991; Sheldon and Tabor, 2009). The elemental and isotopic composition of subsurface B horizons can be used to reconstruct a variety of features including mean annual precipitation (MAP), mean annual temperature (MAT), soil productivity, and atmospheric pCO_2 (Sheldon and Tabor, 2009). Unfortunately, some of those features (e.g. root traces, burrows) reflect complex life that evolved during the Phanerozoic and others (e.g. colour) can change during diagenesis or subsequent exposure. Furthermore, because the climofunctions that are used to reconstruct MAP or MAT are based upon modern vegetated ecosystems, they are inappropriate for use with Precambrian paleosols. This has not prevented some authors (intentionally not cited here) from misusing the Phanerozoic climofunctions, but, at present, readers should be skeptical of any paper that purports to reconstruct MAP or MAT quantitatively from a Precambrian paleosol. Finally, a curious feature of Precambrian paleosols is the dearth of pedogenic carbonate, which is common from the Silurian–present (Sheldon and Tabor, 2009). This, unfortunately, removes one of the primary ways that paleosols can be used to reconstruct pCO_2 (Cerling, 1984). There are at least four possible explanations for this difference between pre- and post-Silurian paleosols, presented in increasing order of likelihood: 1. pre-Silurian ecosystems were too wet or were aseasonal, so carbonate could not be formed; 2. post-Silurian carbonate growth is metabolically mediated by plant and/or microbial processes in the rhizosphere, so carbonate was rare prior to land plants; 3. soil productivity at depth was insufficient to produce enough CO_2 to promote subsurface carbonate growth; or 4. atmospheric pCO_2 was high enough that weathering environments were generally too acidic (pH < 6) for subsurface carbonate growth. Finally, even if Precambrian pedogenic carbonate is identified, because diagenetic alteration of pedogenic carbonate is likely, it will be relatively rare that a reliable reconstruction can be made.

Because of these challenges, other methods have been devised to reconstruct pCO_2 and pO_2 from Precambrian paleosols and to account for diagenetic alteration. Retallack (1992) outlined field criteria for identifying Precambrian paleosols, Maynard (1992) identified key geochemical observations from modern soils and postulated how they differed for Precambrian paleosols, and Rye and Holland (1998) proposed a set of six conditions that need to be met for the chemistry of a paleosol to be considered reflective of the original pedogenic processes: 1. composition different to the underlying parent material; 2. vertical changes in texture through the paleosol profile; 3. vertical changes in mineralogy through the paleosol profile; 4. vertical changes in chemistry through the paleosol profile; 5. evidence for soft-sediment deformation or processes; and 6. sub-greenschist facies metamorphic grade. If all of those conditions are met, then a number of different approaches exist for reconstructing pCO_2 (Figure 3, 5) and pO_2 (Figures 4–5) from Precambrian paleosols as outlined in the following. Strengths and weaknesses of each approach are summarized in Table 1.

One key point is that the approaches to understanding both pCO_2 and pO_2 from Precambrian paleosols encompass a range of model types that include purely observational models that are semi- or non-quantitative in terms of specific gas pressures (e.g. Fe^{3+}/Fe^{2+} ratio), forward modelling (e.g. Alfimova et al., 2014), and inverse modelling (e.g. Sheldon, 2006; Kanzaki and Murakami, 2015).

2.1 Methods for Reconstructing CO$_2$ Levels

Four different methods (Table 1) have been proposed for reconstructing atmospheric pCO_2 using paleosols: 1. steady-state oxidant demand versus acid demand (Eq. 1), 2. mineral equilibrium of co-existing phases, 3. mass-balance modelling of pedogenesis, and 4. mineral dissolution kinetics of pedogenesis. Generally speaking, regardless of which approach is taken, reconstructed pCO_2 during the Precambrian is consistently many times greater than pre-Industrial atmospheric levels (PAL). Among the three methods that have not yet been discussed, the third and fourth can be considered to provide reasonable constraints while the second is problematic for a number of reasons, but all will be discussed briefly. Equation 1 has shown an example of the first method and other papers using this approach are listed in Table 1.

2.1.1 Mineral Equilibrium

The principle behind reconstructing pCO_2 (or pO_2 for that matter) based upon mineral equilibrium is that if coexisting mineral phases were originally formed in equilibrium with one another, then that coexistence reflects the environmental conditions at the time of their formation. All of the previous attempts based upon paleosols (Rye and Holland, 1995), weathering rinds on river gravel clasts (Hessler et al., 2004), and BIFs (Ohmoto et al., 2004) have been based upon the presence or absence of carbonate minerals, typically siderite ($FeCO_3$), or Fe(II)-layered silicate minerals. For example, Hessler et al. (2004) considered the equilibrium between siderite and greenalite ($Fe_3Si_2O_5(OH)_4$) according to the following reaction:

$$3FeCO_3 + 2SiO_{2(aq)} + 2H_2O \Leftrightarrow Fe_3Si_2O_5(OH)_4 \qquad (2)$$

and inferred, based upon the absence of greenalite, that pCO_2 at 3.2 Ga had to be above ~8x PAL for a reaction temperature of 25 °C. In theory, this approach is sound, but there are many issues (Table 1; summarized in greater detail in Sheldon (2006)) with this approach: 1. It assumes that the present-day mineral assemblage is the same as it was at the time of formation, even though many of the rocks have been deeply buried or obviously subjected to diagenesis; 2. It assumes perfectly stoichiometric minerals (e.g. pure siderite versus a mix of siderite and other carbonates); 3. It assumes the temperature is well known or can be estimated relatively precisely, but uncertainties of even 10 °C can dramatically change the results; 4. Most of the existing attempts rely on minerals such as greenalite that represent the metamorphic assemblage and which do not form at Earth's surface conditions; and 5. It assumes that the entire complex process of weathering can be reasonably captured by a single mineral-pair equation. Given all of those limitations, this method can be considered semi-quantitative at best.

2.1.2 Mass Balance

A third approach to reconstructing pCO_2 from paleosols is using mass-balance calculations that characterize acid consumption during pedogenesis. A full description of this modelling approach and a sensitivity analysis of the individual parameters that need to be assumed can be found in Sheldon (2006, 2013), but the basic principle is that the total amount of elemental loss relative to the parent material must be a function of the amount of acid (in the form of CO_2) consumed during pedogenesis, scaled for the length of time over which the paleosol formed. There are three potential sources of acid creation/consumption: 1. acid delivered by rain (X_{rain}), 2. acid delivered by direct diffusion from the atmosphere ($X_{diffusion}$), and 3. acid delivered from biospheric production ($X_{biology}$). Schematically, this is given as:

$$\frac{M}{T} = X_{rain} + X_{diffusion} + X_{biology} \approx pCO_2 \left[\frac{K_{CO_2}r}{10^3} + \kappa \frac{D_{CO_2}a}{L} + f(biology) \right]$$

$$(3)$$

where M/T is the weathering flux (mol cm^{-2}yr^{-1}), K_{CO_2} is the Henry's Law constant for CO_2, r is the rainfall rate (cm yr^{-1}), D_{CO_2} is the diffusion constant for CO_2 in air, a is the ratio of diffusion in soil divided by diffusion in air for CO_2, κ is a constant that relates the ratio of seconds in a year to the number of cm^3 per mole of gas at standard temperature and pressure, and L is the depth to the water table, which for bedrock-parented paleosols, is equivalent to the decompacted thickness of the paleosol. In general, the parameter *f(biology)*, which represents some unknown function for the role of biology, is considered to be essentially zero. While it is unlikely that biological productivity was non-existent in the Precambrian, without a fossil record in most paleosols, this number is very difficult to constrain. Even in cases where microbial or fungal life was present, it was likely concentrated in the upper part of paleosols and sparse or non-existent with depth, which suggests that *f(biology)* $\ll X_{rain}$ or $X_{diffusion}$. The primary strengths of this approach are that it is relatively straightforward to quantify uncertainties, it gives estimates that are internally consistent for 2.7 Ga (Driese et al., 2011), 2.2 Ga (Sheldon, 2006), and 1.1 Ga (Mitchell and Sheldon, 2010; Sheldon, 2013) when there are multiple coexisting paleosols, and it gives results that are consistent with independent estimates from organic fossils and stromato-lites (Figure 3). A final line of support for this model is that the degree of chemical weathering observed in both marine deltaic sediments (i.e. weathered surface sediments; Mitchell and Sheldon, 2016) and other paleosols not used to recon-struct pCO_2 tracks the pCO_2 history that can be reconstructed through the late Paleoproterozoic to early Phanerozoic (Figure 3). The primary weaknesses of this model are that it works best for bedrock-parented paleosols where L can be measured rather than assumed, and that r and T have to be assumed and impart significant uncertainty to the estimates of a factor of 3 (Table 1).

For example, Rybacki et al. (2016) applied the method to a 2.06 Ga weathering profile where L and T were essentially unknowns and did an extensive sensitivity analysis of those parameters. While their best-constrained pCO_2 estimates were in line with penecontemporaneous estimates, they noted that the range of possible error became much larger than when L or T is well constrained. Accordingly, they suggest cases where this method will not be applicable.

2.1.3 Dissolution Kinetics

The final approach to reconstructing pCO_2 from paleosols is through the use of mass dissolution kinetics. A full description of this modelling approach and a sensitivity analysis of the various parameters can be found in Kanzaki and Murakami (2015). In particular, anyone interested in this approach is directed to their extensive supplementary documentation, which outlines the rationale behind their parameterization in great detail. Briefly though, their method follows these steps: 1. calculate porewater anion and cation concentrations at the time of weathering, 2. compare charge balance of calculated anion and cations with carbonate species and dissolved CO_2, and 3. calculate pCO_2 by assuming equilibrium at a given temperature using a pH value derived from

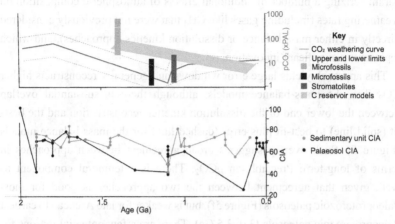

Figure 3 Compiled *p*CO₂ constraints for 2.0 to 0.5 Ga from paleosols and other approaches. Panels show: (A) *p*CO₂ reconstructed from paleosols, C reservoir modelling, microfossils, and stromatolites (references are in the *Supplement* at www.cambridge.org/sheldon); (B) CIA (chemical index of alteration) values from marine shales (Mitchell and Sheldon, 2016) and paleosols (this Element). The paleosol-based reconstruction is generally supported by the other proxy constraints and covers a wider range of time, making it a key constraint on atmospheric composition for the late Paleoproterozoic-Neoproterozoic. CIA values from both paleosols and shales support the *p*CO₂ record as well, with the highest CIA values (i.e. most weather sediments) corresponding to high *p*CO₂ values and the lowest CIA values corresponding to low *p*CO₂ values. A key feature of the all of the records is a shift to low *p*CO₂ values in the late Mesoproterozoic that represent a variation of the 'faint young Sun' paradox and a CH₄ paradox (see Section 3 Discussion).

secondary weathering products (e.g. phyllosilicates). This model can be thought of as a hybrid between the equilibrium and mass-balance approaches, but where reaction kinetics are constrained using model mineral behaviour. The primary strengths of this approach are that it accounts for a wide variety of dissolved and mineral phases and that all of the parameterizations have a sound physical basis from either modern soils or from mineral dissolution experiments. The primary weaknesses are that this approach requires assumptions about a lot of param- eters, and that the end-product *p*CO₂ estimates are still based on mineral equilibrium, with all of the issues associated with those approaches (Table 1). One particular challenge is the long-recognized discrepancy between field and laboratory-derived reaction and weathering rates (White and Brantley, 2003). Kanzaki and Murakami (2018a, 2018b) thus revised their earlier work by parameterizing a number of additional effects of atmospheric composition on weathering rates (including gases like CH_4 that were not previously considered directly in either mass-balance or dissolution kinetics approaches), and which represents a significant innovation.

This approach yields large error windows and generally reconstructs higher CO_2 than the mass-balance models, although there is substantial overlap between the lower end of the dissolution kinetics reconstruction and the best-fit (solid line) to best-fit plus error (dashed line) for the mass-balance models (Figure 5). Such overlap suggests broad agreement between approaches in terms of long-term Precambrian pCO_2. There is a temporal component as well, given that agreement between the two approaches is good for most Paleoproterozoic paleosols (Figure 5f), but is weaker for late Archean to earliest Paleoproterozoic paleosols (3.0–2.5 Ga). There are a few potential reasons for this that relate to each of the approaches. Formation time is particularly poorly constrained as you go back in time, which could lead to lower values for the mass-balance approach and the dissolution kinetic approaches rely, in part, on parameters based upon (or validated with) modern soils or weathering experiments and it has long been known that laboratory experiments result in faster weathering rates than natural systems (e.g. White and Brantley, 2003). At this point, it may be most fruitful to think of the mass-balance approaches as constraining the low-middle part of the plausible range and the dissolution kinetics approaches to define the upper bound of what is plausible. Nonetheless, given the range of error for each approach, it is encouraging that their error ranges overlap for all but *c.* 2.5 Ga ago. Addressing the discrepancy for that critical time (i.e. the earliest Paleoproterozoic) will be a fruitful area for future work.

A final recent approach can be found in Alfimova et al. (2014), who proposed a forward model that includes some common features with all of the equilibrium, mass-balance, and dissolution kinetics approaches. We have placed it primarily with the mass-balance approaches in Table 1 because it relies on modelling water–rock interactions and comparing those model results with measured elemental gains and losses. Alfimova et al.'s modelling involved a range of MAP, pCO_2, and temperature conditions and paleosols ranging in age from 2.8–2.0 Ga. In general, they found that the measured paleosol chemistry was consistent with modelling scenarios characterized by MAP < 1,000 mm yr^{-1}, temperatures of 25 °C or less, and pCO_2 < 25× PAL (Figure 5). These results are consistent with the best-fit to best-fit with lower uncertainty results from the mass-balance models, but they are generally inconsistent with all but the lowest values from dissolution kinetics models, though, as noted, mass-balance approaches may define the lower bound and dissolution kinetics the upper bound of pCO_2. The major challenge to future applications of the Alfimova et al. (2014) model is that, for example, with different parent materials, the model would need to be reconstituted with all new parameters rather than simply reapplied to a new dataset.

2.2 Methods for Reconstructing O_2 Levels

Prior to the discovery of non-mass-dependent fractionation of sulfur (colloquially MIF-S) in sedimentary rocks that predated the GOE and which can only occur at very low pO_2 (Farquhar et al., 2000), two types of continental sedimentary rocks provided some of the primary constraints on the timing of the GOE (e.g. Cloud, 1968): 1. detrital pyrite and uraninite in fluvial rocks, and 2. 'red beds' and paleosol colour and mineralogy (e.g. Figure 1). While MIF-S revolutionized our understand of atmospheric pO_2 record based on marine rocks (reviewed in Farquhar and Wing, 2003), primary sulfide or sulfate minerals are extremely rare in paleosols, so there is only a limited record of terrestrial MIF-S (e.g. Maynard et al., 2013). As a result, five different methods (Table 1) have been proposed for reconstructing atmospheric pO_2 using paleosols: 1. steady-state oxidant demand versus acid demand (Eq. 1); 2. Fe chemistry of Fe^{3+} versus Fe^{2+}; 3. Mass-balance modelling of pedogenesis; 4. mineral oxidation kinetics of pedogenesis; and 5. Cr isotope geochemistry of paleosols or associated erosional products deposited in marine settings. Generally speaking, regardless of which approach is taken, reconstructed pO_2 during the Precambrian is consistently much lower than Phanerozoic levels. Among the five methods, (1) can give quantitative results if pCO_2 is known independently or can be estimated with some confidence (Figure 2), (2) and (3) give a semi-quantitative pO_2 framework, and (4) and (5) can be considered to provide the most quantitative constraints on pO_2.

2.2.1 Fe Geochemistry

Under low pO_2 (typically assumed to be 1% PAL), Fe^{2+} can be dissolved and transported in solution whereas under higher pO_2 Fe is immobile and should all be oxidized to Fe^{3+}. As a result, early work on the Earth's pO_2 took the advent of 'red beds' – a term which unfortunately was used non-genetically – to represent the onset of oxidizing conditions at Earth's surface (Cloud, 1968). Holland (1984) took this a step further by suggesting that paleosols could be used in the same way and proposed that the balance of Fe^{3+} and Fe^{2+} using bedrock-parented paleosols would be particularly valuable because it would be possible to compare the weathering products with the parent value to see whether one redox state or the other was favoured. His work and the work of others found an increase in Fe^{3+}/Fe^{2+} in paleosols between 2.5 and 2.0 Ga ago (Figure 4) that was roughly contemporaneous with a variety of other redox indicators such as detrital uraninites, U contents of black shales, and the disappearance of Superior-style BIFs to indicate the GOE (e.g. summarized in Holland, 1999). Based upon a variety of 'back of the envelope' calculations, Holland (1984, 1999) suggested that this change (i.e. three–sixfold increase in Fe^{3+} to Fe^{2+}; Figure 4) corresponded to an increase of pO_2 to > 1% PAL (i.e. $2x10^{-3}$ bars). This explanation was widely accepted in the community, except by a former Holland PhD student, Hiroshi Ohmoto. Ohmoto used data from some of the same paleosols, normalized to Ti, to conclude that the Fe in the paleosols had instead been significantly oxidized throughout Earth's history, and that pO_2 was > 50% PAL (i.e. 0.1 bars) from 4 Ga onward (Ohmoto, 1996, 1997). The logic is sound in both approaches, so how did the authors come to such dramatically different conclusions?

Holland (1984, 1999) eventually proposed the steady-state framework (Eq. 1) to address some of the issues with using a straight elemental ratio but nonetheless looked at soil formation as akin to ore formation by remineralization. In other words, he did not consider any volume or density loss during weathering, or differences in the initial concentration of the parent rock type that was being weathered (this final point is discussed in some of Holland's paper such as Holland et al. (1989), but not others). Ohmoto (1996) tried to account for differences in initial parent rock chemistry by normalizing Fe^{3+} and Fe^{2+} to the (generally) immobile element Ti, but as a fellow ore-deposit geologist, also failed to account for volume loss or density change, both of which are commonly observed features of soil formation under virtually any weathering conditions (e.g. reviewed in Sheldon and Tabor, 2009). In addition, both Holland and Ohmoto focused primarily on the GOE and on a small subset of (mostly) Precambrian paleosols without seeing if their respective perspectives on Fe geochemistry were consistent with younger parts of the geologic record.

We have expanded the Fe^{3+}/Fe^{2+} dataset to consider the whole period from 3 Ga–present, and while there is a shift in Fe^{3+}/Fe^{2+} at the GOE, the record is largely stable for 2 Ga after that and only shows a second shift in the ratio well into the Phanerozoic (Figure 3), and only for some paleosols. Indeed, a number of Cenozoic paleosols would indicate an O_2-free atmosphere using either of Holland or Ohmoto's interpretive frameworks. Most workers (e.g. reviewed in Lyons et al., 2014) believe that there was a second oxygenation event in the Neoproterozoic that either accompanied or allowed for the rise of multicellular life (e.g. Planavsky et al., 2018), and that shift is also not indicated by a second rise in Fe^{3+}/Fe^{2+}. Thus, on its own, the Fe^{3+}/Fe^{2+} ratio provides only weak evidence of pO_2. We propose that accounting for volume, density, and parent material changes may provide a more nuanced view of Fe oxidation that may eventually provide more quantitative constraints. For example, in Figure 1, the differences between the Fe^{3+}/Fe^{2+} as a raw ratio, accounting for mass balance with an immobile element (Ti), and accounting for density/volume changes are given.

2.2.2 Mass Balance

A second semi-quantitative approach comes from a more nuanced mass-balance method that compares the mobility of transition metals during pedogenesis to immobile elements like Ti, Zr, or Nb, and which accounts for volume change by incorporating measured and reconstructed soil and parent-material bulk densities. Examples of this approach can be found in a number of papers by Steven Driese (Driese, 2004; Driese et al., 2007; Driese et al., 2011), which also include comparisons between mass-balance behaviour of paleosols and comparable modern soils formed under oxygenated conditions. In particular, one of the advantages of this approach is that rather than relying on Fe alone, a variety of redox-sensitive metals can be considered and the differences between their behaviour can be used to provide further constraints on pO_2. For example, Fe^{2+} is mobile up to 1% PAL pO_2 (e.g. Holland, 1984) and then is oxidized to Fe^{3+}, whereas Mn^{2+} is oxidized and immobile at essentially any pO_2 if the pH is relatively alkaline (> 7). Thus, a soil with mobile Fe^{2+} and Mn^{2+} was essentially anoxic whereas one with mobile Fe^{2+} and immobile Mn^{3+} or Mn^{4+} could constrain pO_2 to between anoxic and 1% PAL pO_2. A comprehensive example of this approach can be found in Driese et al. (2011) who used variable transition metal mobility in two penecontemporaneous 2.69 Ga old paleosol profiles with different parent materials to constrain pO_2 and pCO_2 and to make inferences about which metals were organically complexed or microbially mediated.

2.2.3 Oxidation Kinetics

A more quantitative approach to reconstructing pO_2 from paleosols has been proposed by Murakami et al. (2011), who used Fe^{3+}/Fe^{2+} and oxidation kinetics (Table 1) to reconstruct pO_2 between 2.5 and 1.85 Ga ago. Murakami et al. (2011) accounted for volume change during weathering by considering compaction during burial and considered different pathways for Fe^{2+} derived from the paleosol's parent material including dissolution and transport as Fe^{2+}, and oxidation to Fe^{3+} and retention in Fe-bearing minerals. Critically, the estimation of pO_2 requires either estimation or an independent constraint for both pCO_2 and temperature. Murakami et al. (2011) use Kasting's (1993) CO_2-only greenhouse solution to the 'faint young Sun'

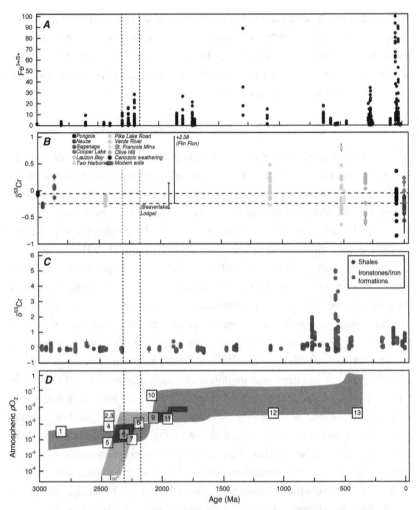

Figure 4 – Compiled pO_2 constraints from paleosols for 3 Ga to the present. Panels show (A) Fe^{3+}/Fe^{2+} compiled from the literature (*this Element*); full references in the *Supplement*, (B) $\delta^{53}Cr$ values from paleosols from Colwyn et al. (2019), (C) $\delta^{53}Cr$ values from marine shales and ironstones; full references in the *Supplement*, and (D) reconstructed pO_2 using steady-state (blue; Rye and Holland, 1998) and oxidation kinetics (grey; Yokota et al., 2013) approaches. Dashed vertical lines represent the period of the loss of non-mass-dependent isotope fractionation of S (Farquhar et al., 2000; Bekker et al., 2004). Numbers correspond to: 1. Mt. Roe #1, #2; 2. Denison/Stanleigh; 3. Quirke II, Cooper Lake; 4. Pronto/NAN; 5. Lauzon Bay; 6. Hokkalampi; 7. Ville Marie; 8. Hekpoort; 9. Drakenstein; 10. Wolhaarkop; 11. Flin Flon; 12. Sturgeon Falls; and 13. Arisaig paleosols where full references can be found Rye and Holland

paradox to constrain pCO_2. Those values for pCO_2 are considerably higher than either fossil-based or paleosol mass-balance-based pCO_2 estimates, so this remains a critical parameter in this approach that requires refinement. Murakami et al. (2011) present a number of sensitivity analyses to examine their assumptions for pCO_2 and temperature, and ultimately conclude that their results were not highly dependent on their initial assumptions. Those authors add an additional line of validation based upon Mn oxidation and consider a number of other potentially complicating factors in their supplemental discussion. In general, this approach reconstructs somewhat lower pO_2 prior to the GOE than other approaches and somewhat higher pO_2 during the GOE itself, but with substantial overlap with other results (Figure 4). Thus, this approach shows promise for quantitative pO_2 reconstruction, but additional paleosols, including some spanning the second proposed oxygenation event in the Neoproterozoic, are still needed to validate the results fully.

2.2.4 Cr Isotope Geochemistry

The final approach that has been applied recently is the $\delta^{53}Cr$ composition of paleosols (e.g. Crowe et al., 2013). A more comprehensive discussion of Cr isotope approaches to reconstructing pO_2 can be found elsewhere (Crowe et al., 2013; Babechuk et al., 2017; Colwyn et al., 2019), but a brief summary is presented here. Cr isotope fractionation occurs primarily during redox transformations. For example, ^{52}Cr is retained preferentially relative to ^{53}Cr during manganese oxide–mediated oxidation of Cr(III) to Cr (VI), and Cr(VI) is much more soluble; as a result, products of oxidative weathering are typically ^{53}Cr-depleted relative to the crustal inventory. Isotope mass balance is maintained through aqueous transport of ^{53}Cr to the ocean, where marine sedimentary conditions may re-reduce Cr(VI) to Cr(III) for storage in sedimentary rocks. Thus, under relatively high pO_2 (i.e. > 1% PAL), paleosols should have $\delta^{53}Cr$ values that are more lower relative to the crustal inventory and marine rocks should have $\delta^{53}Cr$ values that are higher relative to the crustal inventory (e.g.

Caption for Figure 4 (cont.)

(1998). Yokota et al. (2013) used paleosols 3, 4, 9, and 11, along with Gaborone (~2.15 Ga), which is described further in their paper. For the Rye and Holland (1998) reconstruction, the numbers correspond to their calculated pO_2 and the blue windows correspond to their calculated error window. For Yokota et al. (2013), the grey window represents their full calculated high to low range. NB, at 2.5 Ga, Yokota et al.'s (2013) minimum value is off scale at 10^{-12}.

Crowe et al., 2013). These conditions are met in both paleosol and marine $\delta^{53}Cr$ in the Neoproterozoic (Figure 4), although the two sets of data do not correspond perfectly in time because of the incompleteness of the sedimentary record.

While there are a number of potential pitfalls (e.g. complexation of Cr with organic matter rather than Mn-oxides) to $\delta^{53}Cr$ geochemistry that are discussed in Babechuk et al. (2019) and Colwyn et al. (2019), the current state of our knowledge is promising, and there is potential for expanding both the mineral phases that are being examined and the number of samples being analyzed. In addition, $\delta^{53}Cr$ geochemistry adds a more nuanced view to surface redox evolution. For example, while 1.1 Ga paleosols have high Fe^{3+}/Fe^{2+} values, they also have $\delta^{53}Cr$ values that match the crustal inventory (Figure 4), indicating pO_2 at or below 1% PAL. This prompted Planavsky et al. (2018) to look more closely at Fe^{3+}/Fe^{2+} values while considering parent material geochemistry and they concluded that both Fe^{3+} and Fe^{2+} had actually been lost from the paleosols. Finally, it should be noted that early paleosol $\delta^{53}Cr$ papers (e.g. Crowe et al., 2013) used the measured $\delta^{53}Cr$ values but did not consider either volume loss or density change. Colwyn et al. (2019) considered these factors and, in so doing, some previously published values that had been taken as evidence for pre-GOE elevated pO_2 were reduced to the crustal inventory and no longer provide evidence for elevated pO_2 (details therein). A final challenge for whole Earth $\delta^{53}Cr$ geochemistry is that while the marine and terrestrial Cr reserves should be complementary (i.e. higher values in one should be matched by lower values in the other for a given time period), they are not in many cases (e.g. ca. 1.1 Ga). This is an open area of study, but some potential solutions are related to the total weatherable area during the Precambrian (Lalonde and Konhauser, 2015) or to transport processes between terrestrial weathering and marine sinks (Wei et al., 2020).

3 Discussion

3.1 Paleosol-based pCO_2 and pO_2 through Time

One obvious question is whether paleosol-based pCO_2 and pO_2 records are consistent with other proxy-based records of Precambrian atmospheric composition and with other long-term geologic processes. In Figure 3, a variety of methods for reconstructing pCO_2 from 2.0–0.5 Ga are compared. There is substantial overlap for time periods where multiple proxies can be compared (e.g. 1.2–0.9 Ga ago). For others, more work will be needed in the future. For example, there are not currently any paleosols described between 1.8 and 1.2 Ga ago, so the pCO_2 curve from Sheldon (2013) for that time period is a hypothesis, and the one available microfossil-based result suggests that values may have

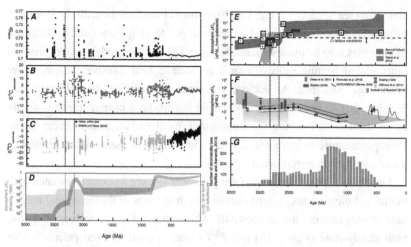

Figure 5 Summary figure for 3 Ga-present comparing paleosol results to other measures of global environmental and biological change. Panels show: (A) $^{87/86}$Sr from marine carbonates; (B) δ^{13}C from marine carbonates; (C) δ^{18}O from marine carbonates; (D) atmospheric pO_2 from modelling (Kasting, 1993) and an array of marine and terrestrial redox proxies compiled by Lyons et al. (2014); (E) pO_2 from paleosols using the steady-state and oxidation kinetics approaches (Table 1), with the horizontal blue line representing constraints from paleosol δ^{53}Cr; (F) pCO_2 from an array of approaches including atmospheric modelling (Kasting, 1993) and paleosol-based reconstructions; and (G) the diversity of stromatolite taxa. Dashed vertical lines represent the period of the loss of non-mass dependent isotope fractionation of S (Farquhar et al., 2000; Bekker et al., 2004). Additional references for data used to compile these figures are given in the *Supplement*.

been higher. In general, the various proxies agree within error and as discussed previously, the observed patterns in surface weathering in shallow marine sediments, floodplain sediments, and paleosols not used in the pCO_2 reconstruction follow the general trends of periods of high CO_2 (and therefore, higher degrees of chemical weathering) identified from the paleosol pCO_2 reconstruction.

From 3.0 Ga to present (Figure 5), paleosol derived estimates of pCO_2 generally fit well with atmospheric modelling based upon maintaining different surface temperatures with a CO_2-H_2O–only greenhouse (i.e. Kasting, 1993). The estimates tend to be most consistent with lower surface temperature model results, but the total radiative forcing needed for higher temperatures could also be realized if other greenhouse gases such as CH_4 or N_2O were relatively abundant (see Section 3.2; e.g. Wordsworth and Pierrehumbert, 2013). Estimates of total necessary CO_2-derived greenhouse gas forcing from those 1-D climate models may also be overestimated, because 3-D climate models that incorporate other global heating processes such as oceanic heat transport require lower greenhouse gas loads to achieve the same temperatures (e.g. Fiorella and Sheldon, 2017). In addition, paleosols do not record a change in pCO_2 during either the GOE or during the Lomagundi–Jatuli carbon isotope event in the Paleoproterozoic (Figure 5B; 2.2–2.0 Ga) that is sometimes interpreted as a massive carbon burial event.

Atmospheric pO_2 constraints from paleosols (Figure 4) generally match both model and metal-redox derived estimates both in terms of the timing and magnitude of pO_2 changes during the GOE. Although the data are somewhat sparse, both steady-state (Figure 5E) and $\delta^{53}Cr$ data (Figure 4) from paleosols also support a prolonged low pO_2 state throughout much of the Proterozoic until a second rise in the Neoproterozoic, which is consistent with records based upon redox-sensitive metals in a range of marine environments including deeper water shales and shallower water ironstones and BIFs (Figures 4–5).

The pCO_2 and pO_2 data can also be compared with other long-term changes in the geologic record. For example, peaks in marine Sr radiogenicity do not correspond with periods of high pCO_2 and there is no relationship between reconstructed pCO_2 changes (e.g. Paleoproterozoic-Mesoproterozoic fall; Figures 3, 5) and Sr isotope changes. As others have noted, this suggests that the Sr record is not recording silicate weathering and CO_2 consumption; instead, it is recording changes in what is being weathered (e.g. Jacobson et al., 2002). Similarly, some workers have suggested that the Archean ocean could have been substantially warmer (> 75 °C) than during more recent times based upon chert and carbonate $\delta^{18}O$ (Figure 5C), however this idea is not supported by multi-isotope studies of cherts (Hren et al., 2009) or $\delta^{18}O$ of phosphates (Blake et al., 2010). Given the consistency of reconstructed pCO_2 levels among various proxies and that all of the different methods indicate relatively modest pCO_2 in contrast to the many bars of pCO_2 that would be necessary to maintain the proposed extreme temperatures, the paleosol record does not support surface conditions that were substantially warmer for at least the past 3 Ga. Instead, the paleosol results are consistent with research suggesting that much of the Precambrian carbonate $\delta^{18}O$ record may have been impacted by diagenetic alteration (Galili et al., 2019). A final observation related to carbon burial is that while the marine $\delta^{13}C$ record is relatively flat throughout the Paleo- and Mesoproterozoic (Holland's 'boring billion'), paleosols and other proxies (Figure 3) indicate declining pCO_2 that reaches a low point in the late Mesoproterozoic before rebounding to higher levels that persist through much of the Neoproterozoic and Phanerozoic. This nadir corresponds roughly with the peak diversity and abundance of stromatolites in both marine and terrestrial settings (Figure 5 G), so it is possible that the two phenomena may be linked through more efficient burial of carbon by microbial communities (Sheldon, 2013). An alternative proposed mechanism is an increase in the efficiency of organic carbon burial that was not accompanied by a change in carbonate carbon burial. Whatever the ultimate cause, the paleosol record has highlighted an interesting problem for future work.

3.2 On the Mesoproterozoic 'CH_4 Paradox'

A final problem that has come out of understanding pCO_2 and pO_2 from paleosols is the late Mesoproterozoic 'CH_4 paradox', which can be thought of as a variant of the 'faint young Sun' paradox. Mitchell and Sheldon (2010) and Sheldon (2013) identified the minimum in pCO_2 that is reached by ~1.1–0.9 Ga ago. In the absence of another greenhouse gas, those relatively low levels of pCO_2 values would have resulted in very cold surface temperatures, yet there is no evidence of glaciation between 60°S and 60°N (Fiorella and Sheldon, 2017). Measurements of $\delta^{53}Cr$ from same paleosols and from contemporaneous marine sediments (Figure 4) suggest very low pO_2. In the present atmosphere, CH_4 is relatively unstable and oxidizes to CO_2 on decadal timescales, but at low pO_2 ($< 10^{-3}$ PAL), CH_4 can potentially accumulate to high levels. As a result, CH_4 has often been invoked as an alternative greenhouse gas both early on in Earth's history to compensate for the 'faint young Sun' and in the late Mesoproterozoic. At the same time, Olson et al. (2016) found that marine CH_4 production was likely limited by low SO_4, which in turn is a result of low pO_2. Thus, low atmospheric pCO_2 combined with low potential marine CH_4 production creates a 'CH_4 paradox' (Olson et al., 2016; Fiorella and Sheldon, 2017; Zhao et al., 2017; Laakso and Schrag, 2019; Hren and Sheldon, 2020). While this issue is far from settled, at least two potential solutions have been proposed that merit future investigation. Fiorella and Sheldon (2017) noted that their global climate model for the Mesoproterozoic could solve the Mesoproterozoic version of the 'faint young Sun' paradox either via CH_4 or via another greenhouse gas such as N_2O. Similarly, Zhao et al. (2017) considered CH_4 fluxes from terrestrial systems and found that with 10% or less coverage of the Earth's surface by wetland settings with methanogens present, CH_4 production from the land could compensate for a lack of CH_4 production from the oceans.

4 Future Prospects

The use of paleosol geochemistry to reconstruct pCO_2 and pO_2 is relatively mature, with a variety of approaches and refinements proposed starting in the

1980s (Table 1). While there are a range of semi-quantitative and quantitative approaches available, further refinements and reductions in uncertainty are possible. For example, while we have highlighted Cr isotopes as a tool for constraining pO_2, there are a variety of other transition metal isotope systems that have been applied primarily to marine systems that could potentially be useful in continental settings as well. For example, Fe, Mo, and Cu all have common stable isotopes and have distributions that can change during redox transformations. Whereas records derived from, for example, marine shales are far removed from the atmosphere and may require a second redox transformation (i.e. similar to Cr in marine settings), paleosols have the potential to record environmental conditions more directly. The mass balance of other metals such as U and V could also prove fruitful as either semi-quantitative or quantitative constraints on pO_2, but need further investigation. While the record of MIF-S from paleosols is currently very limited (i.e. Maynard et al., 2013), discovery of new paleosols with either primary sulfate or sulfide minerals could also provide a critical constraint on pO_2, and similarly, primary pedogenic carbonates could potentially be used with $\Delta^{17}O$ isotopes to constrain either pCO_2 or pO_2 if the other value is known or assumed (e.g. Planavsky et al., 2020).

An additional area of evolving work relevant to thinking about planetary atmospheres in general, is proxies for the total atmospheric pressure and whether it may have been higher or lower than at present. While there is great uncertainty between the methods with some indicating lower total pressure (e.g. Som et al., 2012 but cf. Kavanagh and Goldblatt, 2015) and others indicating potential trade-offs between CO_2 and N_2 (Payne et al., 2020), but possibly high total pressure, the relative importance of both greenhouse gases and O_2 could be very different if the Earth has a dynamic atmospheric pressure history. For example, under some conditions, O_2 acts not just as the redox barometer but also as an additional greenhouse gas (Poulsen et al., 2015). This effect has not yet been considered for the Precambrian.

As discussed, even though Fe chemistry has long been applied, its use as a pO_2 indicator may be improved by re-examining its geochemistry from a soil science rather than an ore deposits perspective. In addition, both the dissolution and oxidation kinetics approaches are based upon the geochemistry of single minerals (e.g. biotite), or of a small number of minerals, so improved empirical weathering experiments of additional minerals or rocks, or theoretical models that incorporate kinetics and mass balance (e.g. Alfimova et al., 2014) could refine our understanding of pCO_2 and pO_2. Although relatively unlikely, the discovery of primary Precambrian pedogenic carbonate would also open another avenue for pCO_2 reconstruction using methods developed for studying Phanerozoic paleosols (Cerling, 1984;

Sheldon and Tabor, 2009). Finally, all of the potential developments to studying Precambrian paleosols on Earth could provide new insights into weathering processes that could be applied to the study of Mars, other rocky bodies in the Solar System, and to the study of exoplanets studied either remotely or perhaps, eventually, directly.

5 Conclusions

The geochemistry of paleosols is one of the critical archives for understanding the evolution of Earth's surface and of pCO_2 and pO_2. In particular, paleosol Fe chemistry has long been considered one of the primary lines of evidence for the GOE. However, the compilation of larger data sets and the use of principles from soil science suggest the need for a more nuanced use of Fe chemistry. More recent innovations in Cr geochemistry provide constraints on pO_2 in the Mesoproterozoic and Neoproterozoic, giving insights into potential drivers for the evolution of multi-cellularity. As with emerging marine redox proxies, the simple three-stage model for the evolution of atmospheric O_2 proves to be an oversimplification in paleosol records, suggesting that further work on redox sensitive transition metals is needed. Paleosols also support relatively low pCO_2 throughout the Precambrian from at least 2.7 Ga to the present based upon most approaches (Figure 5), and those low levels are validated by independent approaches based upon both organic and carbonate micro- and macrofossils. With pCO_2 insufficient to overcome the 'faint young Sun' paradox both in the Archean and Mesoproterozoic, this suggests that other greenhouse gases such as CH_4 or N_2O were likely necessary components of the Precambrian atmosphere, or that we need to consider both marine and terrestrial greenhouse gas production, or that total atmospheric pressure may have varied through time dynamically. Prior to ~2.8 Ga, pCO_2 could have been higher, but estimates will likely come from archives other than paleosols because paleosols of that age are rarely present and when they are, they have typically been metamorphosed beyond greenschist facies. The use of paleosol geochemistry to constrain atmospheric composition during the Precambrian has made substantial progress since it began in the 1980s and has been a key archive for understanding critical global events like the GOE, as well as more subtle problems involving the co-evolution of life and its environment. With recently described and promising emerging techniques, paleosol geochemistry will continue to be one of the key proxies for understanding how the Earth's atmosphere has evolved and may eventually present insights into understanding the evolution of the atmospheres of Solar System planets and beyond.

References

Key References (10 important papers)

Crowe, S. A., Dossing, L. N., Beukes, N. J., et al., 2013. Atmospheric oxygenation three billion years ago. *Nature* 501, 535–538.

Kanzaki, Y., Murakami, T., 2015. Estimates of atmospheric CO_2 in the Neoarchean-Paleoproterozoic from paleosols. *Geochimica et Cosmochimica Acta* 159, 190–219.

Maynard, J. B., 1992. Chemistry of modern soils and a guide to interpreting Precambrian paleosols. *The Journal of Geology* 100, 279–289.

Ohmoto, H., 1996. Evidence in pre-2.2 Ga paleosols for the early evolution of atmospheric oxygen and terrestrial biota. *Geology* 24, 1135–1138.

Retallack, G. J., 1991. Untangling the effects of burial alteration and ancient soil formation. *Annual Reviews of Earth and Planetary Sciences* 19, 183–206.

Retallack, G. J., 1992. How to find a Precambrian paleosol, in Schidlowski, M., Golubic, S., Kimberley, M. M., McKirdy, D. M., Trudinger, P. A. (eds.), Early Organic Evolution and Mineral and Energy Resources. Berlin: Springer, pp. 16–30.

Rye, R., Holland, H.D., 1998. Paleosols and the evolution of atmospheric oxygen: a critical review. *American Journal of Science* 298, 621–672.

Rye, R., Kuo, P. H., Holland, H. D., 1995. Atmospheric carbon dioxide levels before 2.2 billion years ago. *Nature* 378, 603–605.

Sheldon, N. D., 2006. Precambrian paleosols and atmospheric CO_2 levels. *Precambrian Research* 147, 148–155.

Sheldon, N. D., Tabor, N. J., 2009. Quantitative paleoenvironmental and paleoclimatic reconstruction using paleosols. *Earth Science Reviews* 95, 1–52.

Additional References

Alfimova, N. A., Novoselov, A. A., Matrenichev, V. A., de Souza Filho, C. R., 2014. Conditions of subaerial weathering of basalts in the Neoarchean and Paleoproterozoic. *Precambrian Research* 2014, 1–16.

Anbar, A. D., Duan, Y., Lyons, T. W., et al., 2007. A whiff of oxygen before the great oxidation event? *Science* 317, 1903–1906.

Babechuk, M. G., Kleinhanns, I.C., Schoenberg, R., 2017. Chromium geochemistry of the ca. 1.85 Ga Flin Flon paleosols. *Geobiology* 15, 30–50.

Babechuk, M. G., Weimar, N., Kleinhanns, I. C., et al., 2019. Pervasively anoxic surface conditions at the onset of the Great Oxidation Event: new multi-proxy constraints from the Cooper Lake paleosol. *Precambrian Research* 323, 126–163.

Bekker, A., Holland, H. D., Wang, P.-L., et al., 2004. Dating the rise of atmospheric oxygen. *Nature* 427, 117–120.

Blake, R. E., Chang, S. J., Lepland, A., 2010. Phosphate oxygen isotopic evidence for a temperate and biologically active Archaean ocean. *Nature* 464, 1029–1032.

Cloud, P. E., Jr., 1968. Atmospheric and hydrospheric evolution on the primitive Earth. *Science* 160, 729–736.

Cerling, T. E., 1984. The stable isotopic composition of modern soil carbonate and its relationship to climate. *Earth and Planetary Science Letters* 71, 229–240.

Colwyn, D. A., Sheldon, N. D., Maynard, J. B., et al., 2019. A paleosol record of the evolution of Cr redox cycling and evidence for an increase in atmospheric oxygen during the Neoproterozoic. *Geobiology*, doi: 10.1111/gbi.12360

Dimroth, E., Kimberley, M. M., 1976. Precambrian atmospheric oxygen: evidence in the sedimentary distributions of carbon, sulfur, uranium, and iron. *Canadian Journal of Earth Sciences* 13, 1161–1185.

Driese, S. D., 2004. Pedogenic translocation of Fe in modern and ancient Vertisols and implications for interpretations of the Hekpoort paleosol (2.25 Ga). *Journal of Geology* 112, 543–560.

Driese, S. G., Jirsa, M. A., Ren, M., et al., 2011. Neoarchean paleoweathering of tonalite and metabasalt: implications for reconstructions of 2.69 Ga early terrestrial ecosystems and paleoatmospheric chemistry. *Precambrian Research* 189, 1–17.

Driese, S. G., Medaris Jr., L. G., Ren, M., Runkel, A. C., Langford, R. P., 2007. Differentiating pedogenesis from diagenesis in early terrestrial weathering surfaces formed on granitic composition parent materials. *Journal of Geology* 115, 387–406.

Farquhar, J., Bao, H., Thiemens, M., 2000. Atmospheric influence of Earth's earliest sulfur cycle. *Science* 289, 756–758.

Farquhar, J., Wing, B. A., 2003. Multiple sulfur isotopes and the evolution of the atmosphere. *Earth and Planetary Science Letters* 213, 1–13.

Fiorella, R. P., Sheldon, N. D., 2017. Equable end Mesoproterozoic climate in the absence of high CO_2. *Geology* 45, 231–234.

Galili, N., Shemesh, A., Yam, R., et al., 2019. The geologic history of seawater oxygen isotopes from marine iron oxides. *Science* 365, 469–473.

Gay, A. L., Grandstaff, D. E., 1980. Chemistry and mineralogy of Precambrian paleosols at Elliot Lake, Ontario, Canada. *Precambrian Research* 12, 349–373.

Hessler, A. M., Lowe, D. R., Jones, R. L., Bird, D. K., 2004. A lower limit for atmospheric carbon dioxide levels 3.2 billion years ago. *Nature* 428, 736–738.

Holland, H. D., 1999. When did the Earth's atmosphere become oxic? A Reply. *The Geochemical News* 100, 20–22

Holland, H. D., 1984. The Chemical Evolution of the Atmosphere and Oceans. Princeton, NJ: Princeton University Press.

Holland, H. D., Feakes, C. R., Zbinden, E. A., 1989. The Flin Flon paleosol and the composition of the atmosphere 1.8 bybp. *American Journal of Science* 289, 362–389.

Holland, H. D., Zbinden, E. A., 1988. Paleosols and evolution of the atmosphere: part I, in Lerman, A., Meybeck, M. (eds.) Physical and Chemical Weathering in Geochemical Cycles. Dordecht: Reidel, pp. 61–82.

Hren, M. T., Sheldon, N. D., 2020. Terrestrial microbialites provide constraints on the Mesoproterozoic atmosphere. *The Depositional Record* 6, 4–20. doi: 10.1002/dep2.79

Hren, M. T., Tice, M. M., Chamberlain, C. P., 2009. Oxygen and hydrogen isotope evidence for a temperate climate 3.42 billion years ago. *Nature* 462, 205–208.

Jacobson, A., Blum, J. D., Walter, L. M., 2002. Reconciling the elemental and Sr isotope composition of Himalayan weathering fluxes: insights from the carbonate geochemistry of stream waters. *Earth and Planetary Science Letters* 66, 3417–3429.

Kanzaki, Y., Murakami, T., 2018a. Effects of atmospheric composition on apparent activation energy of silicate weathering: I. Model formulation. *Geochimica et Cosmochimica Acta* 233, 159–186.

Kanzaki, Y., Murakami, T., 2018b. Effects of atmospheric composition on apparent activation energy of silicate weathering: II. Implications for evolution of atmospheric CO_2 in the Precambrian. *Geochimica et Cosmochimica Acta* 240, 314–330.

Kasting, J. F., 1993. Earth's early atmosphere. *Science* 259, 920–926.

Kavanagh, L., Goldblatt, C., 2015. Using raindrops to constrain past atmospheric density. *Earth and Planetary Science Letters* 413, 51–58.

Kenrick, P., Crane, P. R., 1997. The origin and early evolution of plants on land. *Nature* 389, 33–39.

Laakso, T. A., Schrag, D. P., 2019. Methane in the Precambrian atmosphere. *Earth and Planetary Science Letters* 522, 48–54.

Lalonde, S. V., Konhauser, K. O., 2015. Benthic perspective on Earth's oldest evidence for oxygenic phtosynthesis. *Proceedings of the National Academy of Sciences (USA)* 112, 995–1000.

Lyons, T. W., Reinhard, C. T., Planavsky, N. J., 2014. The rise of oxygen in Earth's early ocean and atmosphere. *Nature* 506, 307–315.

Maynard, J. B., Sutherland, S. J., Rumble III, D., Bekker, A., 2013. Mass-independently fractionated sulfur in paleosols: a large reservoir of negative Δ^{33}S. *Chemical Geology* 362, 74–81.

Mitchell, R. L., Sheldon, N. D., 2010. The ~1100 Ma Sturgeon Falls paleosol revisited: implications for Mesoproterozoic weathering environments and atmospheric CO_2 levels. *Precambrian Research* 183, 738–748.

Mitchell, R. L., Sheldon, N. D., 2016. Sedimentary provenance and weathering processes in the 1.1 Ga Midcontinental Rift of the Keewenaw Peninsula, Michigan, USA. *Precambrian Research* 275, 225–240.

Murakami, T., Matsuura, K., Kanzaki, Y., 2016. Behaviors of trace elements in Neoarchean and Paleoproterozoic paleosols: implications for atmospheric oxygen and continental oxidative weathering. *Geochemica et Cosmochimica Acta* 192, 203–219.

Murakami, T., Sreenivas, B., Sharma, S. D., Sugimori, H., 2011. Quantification of atmospheric oxygen levels during the Paleoproterozoic using paleosol compositions and iron oxidation kinetics. *Geochimica et Cosmochimica Acta* 75, 3982–4004.

Ohmoto, H., 1997. When did the Earth's atmosphere become oxic? *The Geochemical News* 93, 12–13, 26–27.

Ohmoto, H., Watanabe, Y., Kumazawa, K., 2004. Evidence from massive siderite beds for a CO2-rich atmosphere before ~1.8 billions years ago. *Nature* 429, 395–399.

Olson, S. L., Reinhard, C. T., Lyons, T. W., 2016. Limited role for methane in the mid-Proterozoic greenhouse. *Proceedings of the National Academy of Science* 113, 11447–11552.

Payne, R. C., Brownlee, D., Kasting, J. F., 2020. Oxidized micrometeorites suggest either high pCO_2 or low pN_2 during the Neoarchean. *Proceedings of the National Academy of Sciences* 117, 1360–1366.

Pinto, J. P., Holland, H. D., 1988. Paleosols and the evolution of the atmosphere: part II, in, Reinhardt, J., Sigleo, W., (eds.) *Paleosols and Weathering through Geologic Time*. Geological Society of America Special Paper 216, pp.21–34.

Planavsky, N. J., Cole, D. B., Isson, T. T., et al., 2018. A case for low oxygen during Earth's middle history. *Emerging Topics in Life Sciences* 2, 149–159.

Planavsky, N. J., Reinhard, C. T., Isson, T. T., Ozaki, K., Crockford, P. W., 2020. Oxygen isotope fractionations in Mid-Proterozoic sediments: *evidence for a low-oxygen atmosphere? Astrobiology* 20 (5), doi: http://doi.org/101.1089/ast.2019.2060

Poulsen, C. J., Tabor, C., White, J. D., 2015. Long-term climate forcing by atmospheric oxygen concentration. *Science* 348, 1238–1241.

Retallack, G. J., 2013. A short history and long future for paleopedology, in Driese, S. G., Nordt, L. (eds.), *New Frontiers in Paleopedology and Terrestrial Paleoclimatology*. SEPM Special Publication 104, pp. 5–16.

Rybacki, K. S., Kump, L. R., Hanski, E. J., Melezhik, V. A., 2016. Weathering during the Great Oxidation Event: Fennoscandai, arctic Russia 2.06 Ga ago. *Precambrian Research* 275, 513–525.

Sheldon, N. D., 2013. Causes and consequences of low atmospheric pCO_2 in the Late Mesoproterozoic. *Chemical Geology* 362, 224–231.

Som, S. M., Catling, D. C., Harnmeijer, J. P., Polivka, P. M., Buick, R., 2012. Air density 2.7 billion years ago limited to less than twice modern levels by fossil raindrop imprints. *Nature* 484, 359–362.

Utsunomiya, S., Murakami, T., Nakada, M., Kasama, T., 2003. Iron oxidation state of a 2.45-Byr-old paleosol developed on mafic volcanics. *Geochemica et Cosmochimica Acta* 67, 213–221.

Wei, W., Klaebe, R., Ling, H-F., Huang, F., Frei, R., 2020. Biogeochemical cycle of chromium isotopes at the modern Earth's surface and its application as a paleo-environmental proxy. *Chemical Geology* 541, article 119570.

White, A. F., Brantley, S. L., 2003. The effect of time on the weathering rates of silicate minerals: why do weathering rates differ in the laboratory and in the field? *Chemical Geology* 202, 479–506.

Wordsworth, R., Pierrehumbert, R., 2013. Hydrogen-nitrogen greenhouse warming in Earth's early atmosphere. *Science* 339, 64–67.

Yokota, K., Kanzaki, Y., Murakami, T., 2013. Weathering model for the quantification of atmospheric oxygen evolution during the Paleoproterozoic. *Geochimica et Cosmochimica Acta* 117, 332–347.

Zbinden, E. A., Holland, H. D., Feakes, C. R., 1988. The Sturgeon Falls paleosol and the composition of the atmosphere 1.1GaBP. *Precambrian Research* 42, 141–163.

Zhao, M., Reinhard, C. T., Planavsky, N., 2017. Terrestrial methane fluxes and Proterozoic climate. *Geology* 46, 139–142.

Acknowledgements

NDS acknowledges previous funding from the NSF and NASA in support of various aspects of this research. RLM acknowledges AIM Facility Funding through EPSRC (EP/M028267/1) and Carl Zeiss Microscopy. RMD acknowledges funding from the American Philosophical Society (Lewis and Clark Fund for Astrobiology). This paper was improved by comments from two anonymous reviewers and editor C. Reinhard.

Cambridge Elements $^{\equiv}$

Elements in Geochemical Tracers in Earth System Science

Timothy Lyons
University of California

Timothy Lyons is a Distinguished Professor of Biogeochemistry in the Department of Earth Sciences at the University of California, Riverside. He is an expert in the use of geochemical tracers for applications in astrobiology, geobiology and Earth history. Professor Lyons leads the 'Alternative Earths' team of the NASA Astrobiology Institute and the Alternative Earths Astrobiology Center at UC Riverside.

Alexandra Turchyn
University of Cambridge

Alexandra Turchyn is a University Reader in Biogeochemistry in the Department of Earth Sciences at the University of Cambridge. Her primary research interests are in isotope geochemistry and the application of geochemistry to interrogate modern and past environments.

Chris Reinhard
Georgia Institute of Technology

Chris Reinhard is an Assistant Professor in the Department of Earth and Atmospheric Sciences at the Georgia Institute of Technology. His research focuses on biogeochemistry and paleoclimatology, and he is an Institutional PI on the 'Alternative Earths' team of the NASA Astrobiology Institute.

About the Series

This innovative series provides authoritative, concise overviews of the many novel isotope and elemental systems that can be used as 'proxies' or 'geochemical tracers' to reconstruct past environments over thousands to millions to billions of years—from the evolving chemistry of the atmosphere and oceans to their cause-and-effect relationships with life.

Covering a wide variety of geochemical tracers, the series reviews each method in terms of the geochemical underpinnings, the promises and pitfalls, and the 'state-of-the-art' and future prospects, providing a dynamic reference resource for graduate students, researchers and scientists in geochemistry, astrobiology, paleontology, paleoceanography and paleoclimatology.

The short, timely, broadly accessible papers provide much-needed primers for a wide audience—highlighting the cutting-edge of both new and established proxies as applied to diverse questions about Earth system evolution over wide-ranging time scales.

Cambridge Elements ☰

Elements in Geochemical Tracers in Earth System Science